Dean the Bean gets a Screen

by Snapp&Redden

illustrated by
Nastya Novozhilova

2021

DreamLoud Publishing
dreamloudkids.com

Printed in the United States of America

Dedicated to our nieces and nephews, with love

Dean was a bean who was healthy and green.
Some called him Snap,
his friends called him String.
He loved to dance and he loved to sing
While jumping each day
on his bean trampoline.

His 3-bedroom pod was especially clean,
And he had a good job making bean gas-oline.
With plenty of pals and plenty of bling,
He could've been happy...
except for one thing.

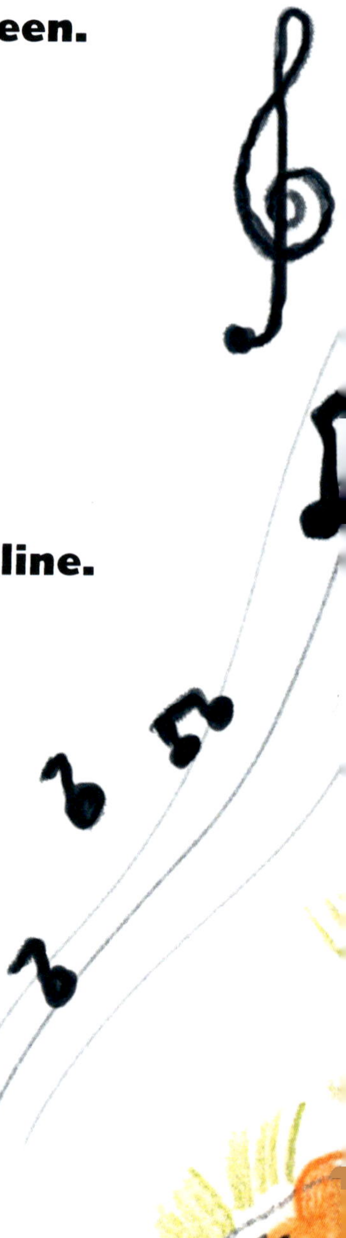

"I'm tremendously bored with being a bean.
I need some adventure – something that's keen!"
"Is there anything new in this vegetable scene?
If so, I must have it, please,
bring it to me!"

Mystical marvel! His plea was received.
Dean's wishes were granted, delivery free!
It fell from the sky in cardboard and foam,
It took the whole wall in his
triple-pod home.

Eyes staring in wonder like black-eyed peas,
He tore through the wrapping
and found a "Bean SCREEN."
"A screen," Dean beamed.
"I've heard about these.
Could this be my key
to being happy and free?"

EWS
for
DAY
SUPER

Dean wasted no time, and grabbing the screen,
He pushed on the button that said, "PUSH ME."
And sud-den-ly...his 3-bedroom pod
seemed more like a dream,
With quick-moving scenes...
and beans talking in streams.
His eyes were transfixed;
"Whoopee-woo" he screamed!
"This screen is like having bean-brain CAFFEINE!"
He checked all the channels and clicked all the apps
He played all the games and skipped his noon nap.

For hours and hours and days and days,
Dean didn't dare move from his screen-amazed daze.
He shut out the world, he stared at the screen.
No baths, no food, no exercise routines.

"Who has time for those useless things?
These pictures and bean-streams are all that I need!"
But within a few days, Dean's cabinets were bare
His stomach was growling for more than just air.

"If only I had a portable screen
To carry to work like a bean magazine.
I mustn't miss out on the latest updates,
Or fail to partake of something
that's great!"

12:00

With a flash of light, and the ring of a bell,
From a cloud of smoke, a small device fell.
It was tiny and shiny and glowed
with a sheen,
And it fit just right
in his green bean jeans.

He yelped out a "Sweeeet!"
and a "Yessiree!"
Got ready for work, eyes glued
to his screen.
"Never again will I be all alone,
I can chat, I can mail, I can talk
on the phone!"

Hi

He watched while he drove to his factory job,
Nearly crashing his car into a corn on the cob!
Driving distracted, he ran through a light,
Missed his left turn and instead took a right.

"I can't stop watching for even a minute,
My favorites I own, the other I'll rent it.
Memes and series and highlights galore,
Every app that I push exposes some more!"

His factory tasks just couldn't compare
With the rivers of info his screen
had to share.

Bran-

"Put that thing down!"
screamed Lima, his boss
It was obvious to all that his boss
was quite cross.
Afraid he'd be fired, Dean watched
from the loo
(A loo is what British beans
call a restroom.)

up??
ok
Today?
gg, How are you? ooh.
the best Photo

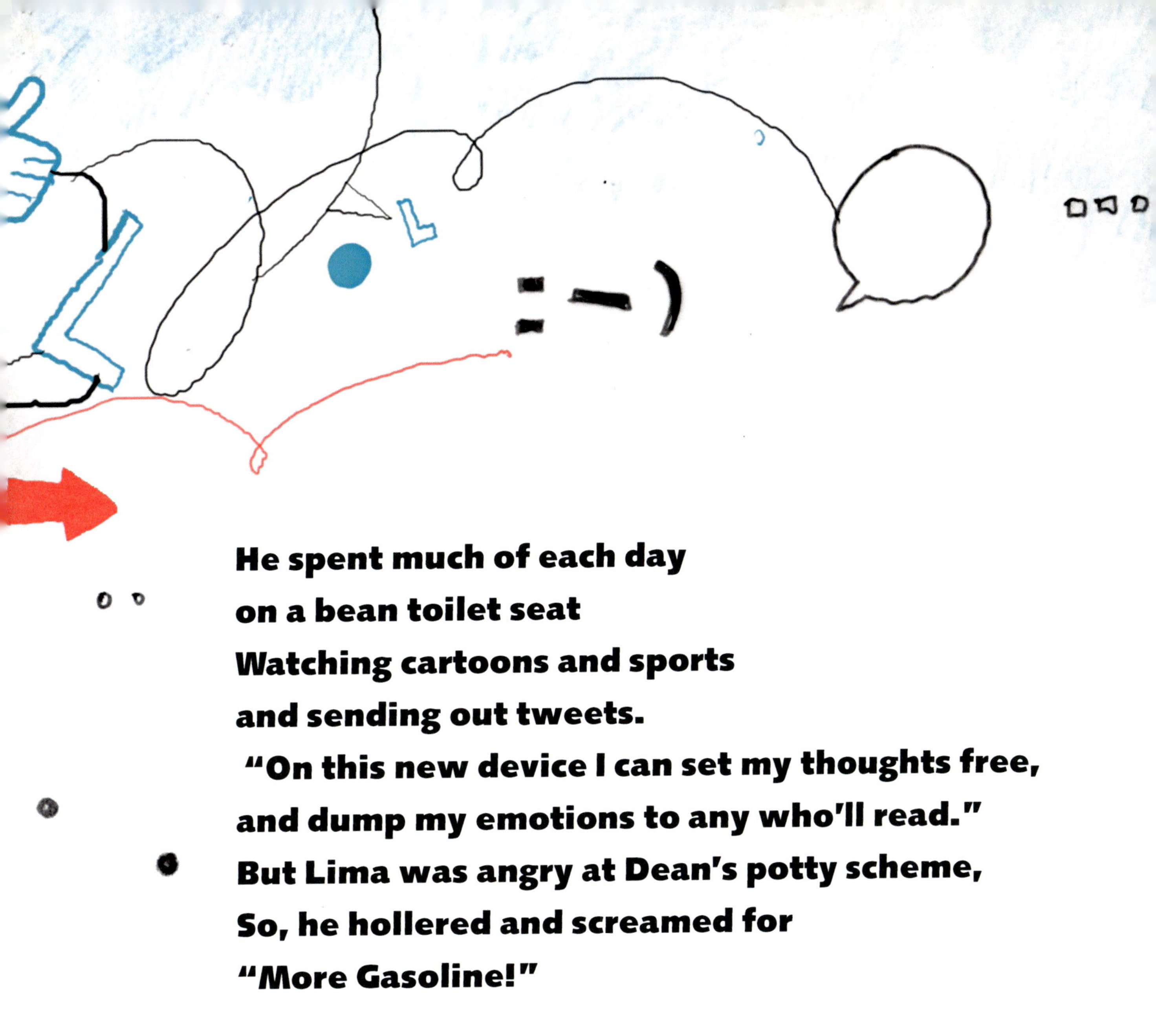

He spent much of each day
on a bean toilet seat
Watching cartoons and sports
and sending out tweets.
"On this new device I can set my thoughts free,
and dump my emotions to any who'll read."
But Lima was angry at Dean's potty scheme,
So, he hollered and screamed for
"More Gasoline!"

Now two years have passed...

FINISH

...and Dean lost his job,
His pod is a mess, and he lives like a slob.
Surrounded by screens he sits in his chair,
His trampoline's dusty, but Dean doesn't care.

His eyes have grown dim,
his mood sadly mellow
His green skin has turned
to a pale shade of yellow.
And my, oh my, does Dean ever stink!
Dean needed a screen in his shower and sink.

It's simple to see how Dean's life went bad
And judge him as being a little bit mad.
"Dean is so stupid," it's easy to say,
"His bean brain is mush, he threw it away!"

But what about you? Do you have a screen?
Does it call you and draw you, like mine does to me?
How much time do you spend on your screen?
Is "Just a little more time,"
what you usually mean?

So, I make this plea to any who'll read,
Screens can be fun, but they aren't all you need.
Be careful to check the things that they do.
Do you control them, or do they control you?

Real people, real work, real playing outside

REAL thinking, real loving real bicycle RIDES!

JOY AND Sadness hugs and close FRIENDS

ARE so much better THAN SCREEN MAKE-PRETEND

!!!

Beware of the screens
that confuse your beliefs
And the products they sell
that give no relief.
They're fine for a time
if you stay really choosy
But Dean would remind you,
the fall is a doozy!

As I say farewell
and wish you good day
I hope you'll allow me one final say...
Turn off your screens
and live out your dreams,
You're a real human being,
not a Do-Nothing Bean!

Bye!

SNAPP & REDDEN

Snapp & Redden are brothers from West Texas who love to laugh and make others smile with their offbeat humor.

They are dedicated to entertaining, encouraging and challenging children of all ages to rise above the obstacles, false expectations, and lies that assault them on a regular basis. The authors' funny, memorable, and often zany outlook, along with their amusing poetic style, are sure to resonate with young and "still feeling young" readers and listeners. The stories give parents an enjoyable teaching platform with which to discuss difficult issues—and if parents aren't careful, they just might learn something, too!

Dean the Bean gets a Screen is their second story to be released but watch for more titles like Tarke the Shark, Toucan't, and a myriad of other heroes they have created for your enjoyment.

If you like the books and want to encourage Snapp & Redden to keep producing, please tell your friends and leave a positive comment with a 5-star rating.

If you'd like to contact them with feedback or ideas for new stories, you can do so through the publishers at DreamLoud Kids via email or Instagram.

Made in United States
North Haven, CT
03 June 2022